STOP GROWTH NOW !!

Arie Kamphorst

ISBN: 1-4392-6210-1
ISBN-13: 9781439262108

Visit www.booksurge.com to order additional copies.

CONTENTS

Preface   v

Part 1: The blog   1

1. Why this blog *(21 July 2008)*   **3**
2. Consolation to Mother Earth *(23 July 2008)*   **5**
3. Paradise regained *(25 July 2008)*   **7**
4. The G8 *(30 July 2008)*   **9**
5. The Doha Conference *(4 August 2008)*   **11**
6. A blessing in disguise *(12 August 2008)*   **13**
7. The illusion of progress *(15 August 2008)*   **15**
8. False prophets *(18 August 2008)*   **17**
9. Wishful thinking *(19 August 2008)*   **19**
10. Ignorance *(25 August 2008)*   **21**
11. The rich and the poor *(5 September 2008)*   **25**
12. Nuclear energy *(7 September 2008)*   **27**
13. Renewable energy sources *(15 September 2008)*   **29**
14. Enough is enough *(16 September 2008)*   **33**
15. Traffic jams *(25 September 2008)*   **35**
16. Witch hunting *(26 September 2008)*   **37**
17. The credit crisis *(13 October 2008)*   **39**
18. Economic stability *(21 October 2008)*   **43**
19. Consumerism and producerism *(29 October 2008)*   **45**
20. Minerals *(13 November 2008)*   **47**
21. Water *(24 November 2008)*   **49**
22. Fish *(30 November 2008)*   **53**

23. Space *(30 November 2008)*   **55**

24. Nature *(30 November 2008)*   **57**

25. Basic instincts *(16 December 2008)*   **59**

Part II: The Postscripts   63

1. Recent developments *(9 July 2009)*   **65**

2. The G8 again *(12 July 2009)*   **69**

3. Fishy business *(20 July 2009)*   **73**

4. Food supply *(22 July 2009)*   **75**

5. Wishful thinking still *(25 July 2009)*   **79**

6. Prospects for the future *(29 July 2009)*   **81**

PREFACE

This book has two distinct parts. Part I was first published on the internet as a blog with twenty-five posts. These posts are presented in this book as twenty-five chapters. They were written from June to December 2008. This means that the blog was started during a period in which several global crises deepened or emerged, namely a climatic change, a deterioration of the natural environment, and shortages of energy, food, water, and minerals.

In the last third of 2008, both a financial crisis and an economic crisis took shape, leading to a decrease in global production and consumption. These crises temporarily overshadowed the shortages of basic commodities. The author expects that the shortages will return as soon as the growth of the world economy is restored, with further attacks on the quality of the environment as a side effect.

Part II of this book consists of six postscripts to the contents of the blog, and was written in July 2009. By that time, the author's observations and warnings had not lost any of their validity and urgency, as all governments were still frantically trying to restore economic growth.

PART I
THE BLOG

In 2008, the Club of Rome, consisting of scientists and top managers from all over the world, celebrated its fortieth anniversary. It published its first report, under the title "Limits to Growth," in 1972. The report contained a warning that continued growth of the world economy and world population would, in the end, lead to the depletion of fossil energy sources and other raw materials, food shortages, and pollution of the environment. For a while, the world was shocked, but this did not last for long. Soon, the emphasis of economists and politicians returned to all out efforts to increase production, consumption, trade, incomes, and job opportunities. The Club of Rome had made the mistake of a too specific prediction of the time-space in which the calamities would take place. This gave producers and consumers alike an opportunity to cast doubt on its report.

After wasting forty years, we are now confronted with the first signs of the predicted shortages. The prices of food, raw materials, and energy are rising. Moreover, it appears that we have been polluting the atmosphere with too high concentrations of carbon dioxide and other greenhouse gasses. Again, producers and consumers are playing down the seriousness of the problems or are suggesting that there are easy solutions. Agricultural scientists try to make us believe that more technology will solve the food crisis, and reactor technologists assure us that the latest types of nuclear reactors are absolutely safe, durable, and clean. Wishful thinkers as human beings are, they are keen to believe them. All that people want in the developed world is to continue their comfortable lives without feeling guilty; in the poor countries, people want to attain the same level of wealth as the inhabitants of the rich countries. Even worse is that politicians and the media tend to participate in this wishful thinking, afraid as they are of a loss of voters, readers, listeners, and viewers.

The most effective contributions to the solution for these world problems would be to stop the growth of the economy, especially in the rich countries, and to stop the growth of the world population. As long as there is a taboo on proposals in this direction, there will be no long-term solution. People who recognise this feel helpless and angry. It is difficult for them to express those feelings, as hardly anybody wants to listen. That is why I have invited the poet Horst Kamparie to publish a poem in this blog. Don't be afraid to read it, as it has a happy end. The poem is reproduced in post 2.

2. Consolation to Mother Earth *(23 July 2008)*

Oh, glorious Mother Earth,
queen of the Universe,
what happens to your grace?
What busy bugs are swarming on your face,
soiling its perfection,
spoiling its complexion?

Oh, gorgeous Mother Earth,
who sent this cruel curse
to you who did not sin,
those creatures crawling on your skin,
destroying with brutality
its health and its vitality?

Don't worry, Mother Earth!
The pest may still get worse
but has not come to stay.
It will soon fade away,
starving due to food depletions,
suffocating in excretions.

Thereafter, Mother Earth,
your face will find rebirth,
beneath new hats of snow and ice

and lucent veils of cloudy skies,
with sparkling blue of seas
and lustrous green of trees,
with subtle hues of rock and sand
and flower shows in pasture land.

And finally, Mother Earth,
belle of the Universe,
you will regain your place
as beauty queen in space,
facing your solar father
and lunar son
in serenity
into eternity.

Horst Kamparie

3. Paradise regained *(25 July 2008)*

Many readers may not have discovered the promised happy end of Horst Kamparie's poem. The poem describes how the earth will survive the attack by the human species, but that is no consolation for those who believe in the future of humanity.

Kamparie explained to me that this is a misinterpretation of what he wrote. It is true that humanity as a pest will fade away, but this does not mean that the human species will disappear altogether. After a period of climatic changes, flooding, starvation, mass migrations, territorial conflicts, and wars, some people may survive somehow. They will get an opportunity to make a fresh start in an environment that will gradually recover from the onslaught of our destructive civilisation. In that environment, the earth may be so thinly populated that there will be no lack of space, food, and shelter. Hence, there will be no property rights, no national boundaries, and no competition for the possession of scarce materials.

In fact, the new world may look like the paradise described in holy books such as the *Koran* and the *Bible*. I am convinced that these descriptions are based on collective memories of a once existing happier world. Of course, it cannot be true that Adam and Eve were the only people living in that world. In that case, their only surviving son Kain would not have succeeded in finding a wife for the conception of their grandson Henoch.

Still, the world must have been so thinly populated that people were hardly aware of each other's existence. There was plenty of nature around them, which provided them with unlimited and renewable amounts of food and other essential commodities.

Paradise must have existed until a rising population density forced people to invent agriculture, animal husbandry, villages, and towns. From then onwards, phenomena like land properties and grazing rights came into existence. These had to be established and defended, which gave rise to personal and communal conflicts. That must have been the cause of people losing their innocence, as is symbolised by Eve eating an apple from the tree of life.

So, the poem has a happy ending after all. Of course, that can be no compensation for all the misery humanity will experience before paradise is regained. Perhaps there are ways to find a new paradise without going through all that suffering. The remaining part of this blog will deal with these possibilities.

R ecently, the club of the eight most powerful industrialised countries held its annual meeting. The concluding declaration of this meeting contained some very important and, at the same time, very surprising statements. First of all, the representatives of the eight countries expressed their grave concern about rapid climatic changes. They recognised that something should be done about this by reducing the emission of carbon dioxide into the atmosphere. To show how very concerned they were, they agreed that CO_2 emissions have to be reduced by 50 percent over a period of forty-two years. This reduction is probably too little and too late. Moreover, the declaration gave no indication of how this reduction should be brought about. Considering the fact that China, India, and many other developing countries refuse to slow down the growth of their economies, the prospects for this ambition are not bright.

Another grave concern of the eight countries is that the growth of the world economy is slowing down, due to the rising shortages and price of oil. It was, therefore, unfortunate that the most important oil producing countries were not invited to the conference. However, it was agreed that pressure should be put on these countries to increase oil production.

Surprisingly, the manifest gave no indication of how CO_2 emissions could be decreased while at the same time production and consumption of oil increased. It is clear that the growth of the world economy, which is happening at present, is considered more important than the climatic change, which will have its most devastating effects only in future. It is time that these priorities are interchanged, and as soon as possible.

5. The Doha Conference *(4 August 2008)*

The conference between economically developed countries and developing countries, which ended on 30 July 2008, reached no agreement on the question of import restrictions and export subsidies. The conference was initiated by the World Trade Organisation (WTO) to promote free trade and liberate world markets.

In view of the problem of climatic change, it may be just as well that no agreements were reached. Globalisation of trade leads to the unnecessary large-scale transport of bulk goods, for which much energy and CO_2 emission is needed. It would be better if countries tried as much as possible to be self-sufficient in the production of necessities for their populations, instead of trying to conquer remote markets.

India was the country that took the lead in opposition against the abolishment of import restrictions. I worked and lived in that country from 1964 to 1971 and I am happy to see that it is continuing a long-standing policy. During my stay in India, the country managed to eradicate famine by increasing its agricultural production. At the same time, it imposed high import duties on Western industrial products. For luxury goods, such as cars, air conditioners, and refrigerators, these import duties were 100 percent or more. In the meantime, its own industrial production was strongly promoted. In doing so,

India has hardly any foreign debts and is now even producing its own cars, aeroplanes, and satellites.

Poor countries should take India as an example for their own development. Instead of producing cattle-fodder and bio-fuels for the rich countries, they should use their soils to become self-supporting in food and other basic requirements. To make this possible, their farmers must be protected against the distribution of cheap food from foreign aid and against the dumping of subsidised agricultural products from rich countries. The present policy of the World Bank in lending money to poor countries has the opposite effect. This money is not used for the improvement of agriculture, but rather for subsidies on food supplies for city dwellers. Much of the money is also diverted to the import of luxury goods by rich people. This is promoted by the fact that the World Bank forces these countries to lift import restrictions.

If the export of subsidised agricultural products to poor countries were stopped, more land would become available in the rich countries for the cultivation of other crops. Then these rich countries could grow the fodder for their cattle and the raw materials for bio-fuels themselves. Imports should be restricted to the essential goods that they cannot produce themselves. A regional economy, as proposed here, would also make the world a more diversified and interesting place.

6. A blessing in disguise *(12 August 2008)*

What the Club of Rome predicted, is happening now. Food and other essential materials are becoming scarce and expensive. Transportation of such goods adds to the increase in the prices, because oil is becoming a scarce and expensive commodity too.

The price increases are disastrous for the poorest people. As they were not responsible for the overconsumption, they should be compensated for this. Provided that this takes place, the oil crisis can be considered a blessing in disguise. Thanks to the rising price of oil, many forms of transport are becoming too expensive, leading to a lower growth of oil consumption and less emission of CO_2. In the USA, people are replacing their big SUVs for smaller European and Japanese cars and air companies are cutting down on the number of inland flights. It is expected in the Netherlands that the annual number of planes using Amsterdam Airport will decrease by ten thousand next year.

All over Europe, fish is becoming expensive and fishing boats are being laid off, which leads to less overfishing of the seas and oceans. We might even see a decrease in the cruel transport of live animals from the Netherlands to remote places all over Europe. For example, at present, pigs are transported in overcrowded trucks from the Netherlands to Italy and returned to their native country in the form of ham from the

Italian province Parma. As Italians are perfectly able to raise pigs and the Dutch can produce very good ham, this practise should be stopped immediately.

Governments should take responsibility in promoting and directing these developments, because their citizens will do their utmost to continue their lives as usual. When the Dutch government imposed an ecotax on air tickets, people went to airports in neighbouring countries to fly to their holiday destinations. When the price of diesel was rising, Portuguese fishermen blocked harbours and French truck drivers did the same with roads, demanding compensation for the rising costs from their governments. At the same time, Dutch car drivers asked for a reduction of the excise on petrol.

Instead of giving in to these demands, governments should confront their citizens with the fact that they must change their life style. People living in the quiet eastern part of the Netherlands and employed in the industrialised western part of the country should be stimulated to live nearer to their work. It would save a lot of fuel, and at the same time, it would reduce the overcrowding of roads by cars going from east to west in the morning and from west to east in the evening. To facilitate this change, the government must stimulate the building of houses in the west and the creation of jobs in the east. Similarly, the daily traffic jams on the highway between Rotterdam and Amsterdam could be ended if people living in Amsterdam and working in Rotterdam swapped houses with those living in Rotterdam and working in Amsterdam. To promote this, companies and other organisations should be forced to abolish travel allowances.

7. The illusion of progress *(15 August 2008)*

When Jesus Christ died almost two thousand years ago, he left behind communes of early Christians. These people tried to live in harmony with each other by agreeing that all members of the commune would "contribute according to their abilities and receive according to their needs." The communes fell apart, because most of the strongest members could not keep up this humane attitude.

In the middle of the nineteenth century, the socialistic movement tried again to adopt this very Christian credo, in reaction to the fact that some people were becoming extremely rich at the expense of big masses of cheap labourers. Surprisingly, this movement was strongly opposed by Christian churches. In the communist countries, where the doctrine was imposed by force, socialism failed. The leaders had to adopt dictatorship, because people were not prepared to use their labour for the common cause.

When socialism failed, capitalism claimed victory. Some writers even proclaimed the end of history, because capitalism would henceforth rule the world. Milton Friedman's Chicago School of Economics, which wanted to limit the influence of governments on the economy, became popular with economists and politicians.

Unfortunately, this neo-liberal capitalistic ideology is based on the same wrong assumption as socialism, that is, the assumption of the altruistic nature of humans. Under this assumption, clever and enterprising people get maximum opportunities to make the world more prosperous and harmonious. In the neo-liberal version of capitalism, the attitude of *laissez faire laissez aller*, which prevailed in the nineteenth century, was revived. The result has been a never-ending increase of production and consumption. Instead of becoming more harmonious, the world has become chaotic. In the USA, where the most extreme form of neo-liberal capitalism is practised, the American dream of a few people is becoming the American nightmare for many.

The present credit crisis shows how capitalists in power have been filling their own pockets at the expense of the common citizens and the environment. And all over the Western world, people are living in luxury at the expense of cheap labour in developing countries. Their conscience is salved by the practice of foreign aid, the most modern version of charity.

The objective of progress should be to attain harmony amongst people as well as between people and nature. At present, the goal seems to be ever increasing economic growth, with the destruction of the environment and nature as a side effect.

8. False prophets *(18 August 2008)*

In the recent past, the economic development in the Western world was strongly influenced by the confidence producers and consumers had in the future. Therefore, politicians and economists adopted the strategy of boosting this confidence. Decreases in economic growth and job opportunities were presented as temporary dips, and increases were attributed to the beginning of new periods of prosperity.

In the near future, economic growth will not be determined anymore by the mood of the people, but instead by the availability of energy, water, minerals, and other raw materials. Under these conditions, boosting of the confidence of producers and consumers must be avoided. To save energy and materials, and to reduce the emission of CO_2, economic growth must be discouraged. There are no signs so far that our leaders are inclined to adopt this strategy. In refusing to do so, they have become false prophets.

A good challenge for the world leaders was offered by the change in oil prices in the recent past. This price had almost doubled during the few past years, due to increasing shortages of oil in the world market. During the past weeks, however, oil prices came down from US$147 to US$115 per barrel. It is tempting to conclude that this reduction marks the end of the "oil crisis," but that would be a serious mistake. The cause of the reduction was a declining demand for oil in the

Western world, due to a reduction of the economic growth to almost zero. In the emerging industrial countries, however, the growth was maintained at its high level of at least 7 percent. Therefore, it must be expected that the price of oil and other raw materials may continue to fluctuate significantly, but at the same time will gradually increase further.

It is important that people be clearly informed about this long-term expectation, so that they can adapt to the new situation of economic decline. In that way, we may still hope for a soft landing for our overheated economy. Otherwise, we will be inviting social unrest, which will ultimately lead to national and international armed conflicts.

9. Wishful thinking *(19 August 2008)*

There are people who tend to ignore the seriousness of the situation. They are the journalists who take advantage of the fact that people are inclined to wishful thinking. Their publications are found mainly in the more popular and successful newspapers and glossy magazines. Popular television programs often adopt the same attitude. Their strategy is to play down the problems or to suggest that there are easy solutions. Many recent newspaper articles, radio programs, and television documentaries reported on cars powered by electricity or hydrogen. All of these journalistic products claimed that such cars were the promise for the future, because they do not emit any CO_2. However, the fact was ignored that CO_2 emission does not occur when electricity or hydrogen is used, but when it is produced.

An interesting case of false prophecy is offered by the Dutch geologist Alexander Kroonenberg. He wrote a book on geology for the general public under the title *The human scale.* In this book, he stated that cold, glacial periods and hot, interglacial periods have occurred repeatedly in the geological past and that the human species cannot have any influence on that. With this message, he contradicted the findings of the climatologists of the International Panel on Climatic Change, who stated that the present rapid global warming is caused by human activity.

The geology professor also comments in his book on the scale of the expected rise of the sea level, which is at present less than a metre in the course of this century, stating that in the past twenty thousand years this level has risen tens of metres. In doing so, he ignores the fact that so many centuries ago, the world was very thinly populated, and people had ample time to retreat in an almost empty hinterland. Now the world is overpopulated with human beings, the majority of whom are living in the fertile river deltas, river plains, and coastal plains. Most of these lowlands are insufficiently protected by dikes. In case of natural disasters, these people cannot retreat to higher land, because it is overpopulated too.

False prophets also are found in organisations trying to sell solutions. Recently, a full professor in reactor technology at one of the leading Dutch technical universities claimed that the latest models of nuclear plants are absolutely clean, durable, and safe. This claim was made while the storage of nuclear waste was still an unsolved problem and while it was known that the availability of uranium is limited.

With respect to safety, a comparison can be made between nuclear reactors and passenger planes. The latter must be very safe too, because they carry hundreds of passengers. Still, it is a fact that many planes crash every year, due to construction failures, lack of maintenance, human mistakes, human aggression, or whims of nature. The same can happen to nuclear plants. The difference between passenger planes and nuclear plants is, however, that in the case of a nuclear accident, many thousands of people may die and large regions may become uninhabitable. Still, the claim of the reactor technologist feeds the wishful thinking of the general public. The acceptance of nuclear power, which used to be very low in the Netherlands, is rising steadily.

10. Ignorance *(25 August 2008)*

Wishful thinking can be described as thoughts based on unfulfilled wishes rather than proven facts. That is why it can thrive best if the "thinker" is not hindered by knowledge and understanding. It is the task of governments and the media to provide this knowledge and understanding. If they fail to do so, false prophets get a chance to feed the people with wrong and harmful information.

In the case of global warming, knowledge and understanding of the processes involved is not very easy to attain. That is why many people think that it can be stopped easily by reducing CO_2 emissions. They do not understand that global warming is a function of the amount of CO_2 that is already in the atmosphere due to emissions in the past. This is why global warming would continue for many decades even if we stopped emitting CO_2 and other greenhouse gasses completely and immediately.

There is another reason to think that it will be very difficult to stop or even retard global warming. This reason is that there are so-called feedbacks in the climate system. An example of such a feedback is the influence of the "melting ice and snow." If, due to global warming, ice and snow were to cover ever-smaller areas during ever-shorter periods, less sun radiation would be reflected into space by these white surfaces. Therefore, more solar energy is absorbed, leading to more global warming, followed by an even faster disappearance of surfaces covered by snow and ice, still more global warming, etcetera. This "loop of reactions" can, in principle, take its own course, even if we stopped CO_2 emissions altogether. In this case, humanity will completely lose control over global warming and the connected climatic changes.

Another feedback is caused by the melting of the permafrost in tundra soils. These are almost permanently frozen peat soils around the periphery of the polar ice. When the tundra melts due to global warming, these soils will emit large quantities of methane or CH_4, which has a fifteen times stronger greenhouse effect than CO_2. Also, this process would trigger a feedback, because more CH_4 in the atmosphere causes more global warming, which in turn causes more melting of permafrost, etcetera.

A potential feedback may also be caused by the methane trapped in ice at the bottom of the oceans. Due to the gradual increase in water temperature, part of this ice will melt, releasing the methane into the water, and from there into the atmosphere.

An interesting anthropogenic feedback is emerging due to the reaction of men to the melting of the ice in the Arctic Sea. All nations bordering this sea are manoeuvring for the best position to exploit the oil and gas fields in it. This exploitation will lead to more emission of CO_2, followed by a faster melting of the ice, a faster exploitation, etcetera.

Many of these feedbacks are not yet introduced in the climate models of the climatologists, because it is difficult to quantify them. It is, however, certain that they do exist. That is probably one of the reasons why global warming is proceeding much faster than was expected by the International Panel on Climatic Change.

The results of a very recent inquiry in the Netherlands with 4642 respondents showed that 66 percent of the people were not worried at all about the effects of global warming on their environment. It may be difficult to bring an understanding of the processes involved to the knowledge of the general public, but it has to be done.

11. The rich and the poor *(5 September 2008)*

Some days ago, the so-called Delta Commission of the Netherlands published a plan, in which it recommended drastic measures to counteract the dangers of the expected rise of the sea level. The Commission expects that the peak discharges of the rivers will increase strongly and that the sea level may rise by 140 cm during this century. To prevent flooding, the dikes along the rivers and the sea must be raised and broadened. To strengthen the dunes, huge amounts of sand must be transported from the bottom of the North Sea to the shoreline. More fresh water must be stored, to counteract the increasing intrusion of salt water, by raising the water level in part of the IJssel Lake by as much as 1.5 metres.

As a result of all these measures, sluices, quays, wharfs, bunds, roads, bridges, buildings, and other structures have to be adapted to the new hydrological situation. The whole operation will cost €1.0 to €1.5 billion (US$1.5 to US$2.0 billion) annually, during a period of one hundred years. For this small country, that is a huge expenditure.

Nevertheless, the report of the Delta Commission has been received with great enthusiasm in many circles. Engineers are keen to show their ability to cope with large and complex problems and politicians are proud of the fact that their small country is again waging war against the water. Engineering firms see the project as a showcase, hoping that they will be invited to carry out similar projects in other rich countries. An example of such a rich country is the USA, where New Orleans has barely escaped a second disaster caused by a hurricane only a week ago. Only few Dutch people are sceptical, because they fear that these measures will play havoc with the natural and cultural heritage of the country.

If these proposals are accepted, the government and the parliament will have admitted for the first time that the climatic change has already become a very serious and almost insolvable problem. The reaction of the country is clearly to save its own skin and retreat behind its dikes. A reduction of the use of fossil energy will become even more difficult, as the battle against the water is nowadays carried out with heavy, oil-slurping machines.

The tragedy of this development is that the populations of poor nations will be left to their own fate. Countries such as Bangladesh, Birma, or Haiti, where cyclones have played havoc before, have no money to build dikes. People on the coral islands in the Pacific Ocean will have no other choice than to evacuate their homes. The populations of North India and South China will suffer too, because the reliable supply of irrigation water will disappear due to the melting of the Himalayan glaciers.

All those inhabitants of the poor countries will have to bear the burden of climatic changes for which they are not responsible. This does not mean, however, that the people in the rich countries will escape the consequences of their past actions altogether. They may have to build high political dikes around their borders, to keep out the streams of economic refugees fleeing from their devastated homelands. Even between European countries, such dikes may have to be built, because recent widespread forest fires indicate that the desertification of southern Europe may have started already.

12. Nuclear energy *(7 September 2008)*

In this blog, some remarks have been made already on the safety of the production of nuclear energy. It was stated that accidents could occur even in the safest nuclear plants due to human failures, human aggression, lack of maintenance, and whims of nature. This is frightening, certainly, if one considers that one nuclear plant would supply only 2 percent of the energy requirement for a small country such as the Netherlands.

Building a nuclear plant of the latest and safest generation would cost around five billion euros and would take about fifteen years. It is clear that this amount of money is available only in the richer countries and that the inhabitants of poor countries will have to continue the use of conventional energy sources. In that way, the economical and technological gaps between the underdeveloped and the overdeveloped worlds will widen even further.

On the other hand, it is just as well that the production of nuclear energy, if it cannot be stopped, will remain in the hands of technologically developed and politically stable countries. Even in these countries, accidents occur from time to time, as was shown by recent leakages of radioactive material from Italian and Japanese reactors. In the Netherlands, a small reactor, which produces isotopes for medical purposes, has been put out of production for the second time, due to problems with the cooling system. According to Greenpeace, accidents occur regularly in French reactors too, but that country manages to cover up most of these occurrences.

In poor countries, the quality of construction and maintenance of nuclear plants will always be in danger, due to a lack of knowledge and funds. There the influence of corruption on building activities is always around the corner. Moreover, governments may try to use nuclear materials and knowledge for the development and production of atom bombs, as might be the case at present in Iran and North Korea. India and Pakistan are already threatening each other with nuclear weapons, and Israel is confronting the Arabic countries with them.

The greatest danger is probably coming from insurgents, who may try to obtain radioactive materials for terroristic attacks. In the recent past, they have shown how inventive they are. They may buy the materials from unstable countries or even steal them from nuclear plants in the developed world. Some years ago, Dr. Khan, from Pakistan, showed how to obtain knowledge of nuclear technology in the Netherlands, and Greenpeace demonstrated in the same country how to infiltrate a nuclear reactor with a large group of armed persons. To prevent these dangers, countries would have to be turned into police states.

The construction of nuclear plants for the production of electricity must be stopped, not only in developing countries, but also in the technologically developed world. There are better, cleaner, more durable, and safer techniques for the production of energy.

13. Renewable energy sources *(15 September 2008)*

If the use of nuclear plants is ruled out, the question arises how our requirements for energy can be covered by other sources. Coal and tar sands are out of the question, because the mining, transport, and purification of these materials require much energy and because these activities harm the health of workers and the beauty of landscapes. Moreover, the use of these fuels pollutes the atmosphere with dust, sulphuric acid, nitric acid, and other contaminants. The most important reason why the use of coal and tar should be avoided, however, is that they produce even more CO_2 per unit energy than oil.

If it would not be possible to switch to renewable energy sources in time, natural gas might temporarily bridge the gap between the production and consumption of energy. Its use produces much less CO_2 than oil, it contains hardly any air pollutants, its exploitation is more healthy for the labourers and the landscape, and its extraction, transport, and purification require less energy. Considerable quantities of this energy source are still available in Russia, Norway, the Netherlands, North Africa, Iraq, and many other countries. It can be used directly for heating and cooking in households and for the production of electricity. In a liquid form, it can also be used as a fuel for transport facilities and mechanical processes.

In the meantime, maximum efforts should be invested in the development of renewable energy sources. There is not much time left for this, as it is expected that the energy requirements of the world will rise by 45 percent by 2030 and that serious oil shortages will arise after 2010.

The most important sustainable, safe, and clean sources are those derived directly or indirectly from solar energy—solar radiation, wind power, and water power. Sceptics of these energy forms state that solar energy cannot be harvested during the night and that the production of wind energy is dependent on the weather conditions. In case of temporal and regional deficiencies, these supplies can, however, be supplemented by energy from hydro-electric power produced at dam sites, from tides and waves, or by the import of solar and aero-electric energy produced elsewhere. For this exchange of energy between different regions, it will be necessary that electricity networks be interconnected over large distances. In that way, China could deliver solar energy to Europe during the Chinese day and vice versa, solar energy for Europe could be produced in the Sahara, and electricity derived from hydrodynamic power could be delivered by Norway to other European countries.

Another possibility is to store solar and aero-electric energy during periods of peak production and release it when this production is deficient. An interesting system for this storage was proposed some years ago in the Netherlands. It consisted of an area surrounded by dams in shallow parts of the IJssel Lake or the North Sea. During periods of high production, the excess electricity could be used to pump water into this bunded area, thus raising the water level in it. During periods of low production, this water could be released again via hydro-electric generators, thus harvesting the energy stored earlier.

In the Netherlands, this system might be incorporated with other plans to raise the dikes and the water level in IJssel Lake. In other countries, this system could be built in any shallow lake or sea.

If one compares wind energy with solar energy, the latter must be considered the best alternative by far. It can be harvested in large parks covered by solar panels, but also by small groups of panels on houses and other buildings. Large modern windmills spoil the landscape and give visual and acoustic hindrance to people living in their surroundings. On land and at sea, their sails form dangerous obstacles for birds. In seas, the vibrations caused by the rotors disorientate sea mammals.

Firms in Europe have started the production of small, very quiet windmills, which can be installed on the roofs of buildings and can produce half of the energy requirement for an average household. The question remains how dangerous these small windmills are for birds. To prevent accidents, the rotor blades may have to be surrounded by screens.

Also the cultivation of crops for the production of so-called bio-fuels must be stopped. These crops occupy agricultural land that would otherwise be used for the production of food and other useful commodities or for the conservation of nature. Moreover, their cultivation will require the use of fertilizers and the transport of bulk materials over large distances.

Electricity will remain by far the best carrier for energy, because it is clean and can be transported easily over large distances. It can be produced directly from all renewable energy sources such as solar, aero-electric, and hydro-electric energy. It can be used directly for different purposes, such as lighting, heating, transport, and mechanical processes. An advantage of electricity as an energy carrier above hydrogen also is that its production, transport, and use are cheaper and more efficient. Moreover, much of the required distribution system for electricity is already available in most countries.

14. Enough is enough *(16 September 2008)*

Protagonists of nuclear energy state that the sustainable and clean energy sources cannot produce enough energy to meet future requirements. Such statements are generally made under the assumption that no special efforts will be made to curb the ever-increasing use of energy. For instance, it is assumed that the economy will continue to grow by at least 3 percent in the rich countries and 7 percent in the developing nations. Also, the continuing growth of the world population is, in their view, a given fact that cannot be influenced.

Other experts claim that it is possible to achieve a reduction of energy consumption by at least 50 percent, merely by stopping the present waste. Many governments have already initiated programs for this purpose, varying from propaganda actions to taxation measures and subsidies. In the Netherlands, these actions are still not taken with great enthusiasm and they are often too limited in scope. When the Dutch government introduced a subsidy on solar boilers about twenty years ago, many citizens bought them, and firms manufacturing and installing them thrived. Fifteen years later, a successive government abolished the subsidy, the firms went bankrupt, and the citizens could not find a replacement for their old boilers.

Other measures that governments can take are increasing taxes on energy-intensive production methods, accompanied by decreasing taxation of labour. This will cause a reduction of the use of energy, but labour productivity, and hence the incomes of families, will decrease as well. We will have to accept this if we really intend to leave behind a hospitable world for our descendents. Of course, a redistribution of incomes would be necessary, to protect the people with the lowest budgets.

If all these measures were insufficient to bring energy consumption in balance with a renewable and sustainable production of energy, a reduction of the world population would be necessary. The most humane way to achieve this is family planning. In rich countries, this is already taking place as a result of the emancipation of women, but in poor countries, the population growth is still too high. The promotion of family planning is difficult there, mainly due to a lack of education, non-emancipation of women, and the availability of facilities for birth control. Also, traditions and religions are important impediments for family planning, as will be discussed later.

15. Traffic jams *(25 September 2008)*

In many parts of the world, traffic congestion has become an ever-increasing problem. In the Netherlands, the most densely populated country in Europe, many measures have been considered to solve this problem. Stopping the growth of the population and of the economy is not one of them. A foundation advocating the gradual reduction of the Dutch population to ten million inhabitants hardly gets any attention, and the government is still frantically trying to save economic growth.

Another possibility to reduce the overcrowding of roads is the introduction of toll collecting. Successive Dutch governments have studied this very effective system, but none of them have dared to introduce it so far. Measures to stimulate people to live closer to their work, work at home, or diversify work hours also are not popular. The government has refrained from imposing such measures and has limited its actions to non-binding agreements with trade and industry. Experience has shown that such agreements are generally not followed in practice. There is also a strong opposition against increases of the excise duties on fuels and of the road tax and sales tax for cars with high fuel consumption.

The only remaining effective solution for the problem is the improvement of public transport, which has been strongly neglected so far. In 2005, there were fewer bus connections than in 1980, due to the recent privatisation of bus lines. The railway system has deteriorated too, due to a lack of maintenance. Also in this case, privatisation has had negative effects.

The latest proposal of the government is to improve both the public transport and the road transport. Unfortunately, no detailed plans for the public transport have been published so far, but the planned improvement of roads has been announced with much publicity. This improvement consists of the widening of roads at bottlenecks. Experts predict that this measure will not be effective for the solution of traffic jams, because the removal of a bottleneck in one place will create a new bottleneck elsewhere, particularly at the entrances to cities. Even if the widening of roads were effective for the immediate disappearance of traffic congestions, it is expected that this effect would be short lived, due to a resulting increase in the number of cars on the roads. It is clear that the government gives preference to the most popular but least effective measures.

16. Witch hunting *(26 September 2008)*

To speed up the widening of roads and other road construction works, the government has proposed a special legislation. Under this legislation, such projects can be approved and executed without lengthy procedures. This means that objections by individuals or non-governmental organisations can be put aside more easily. During his announcement of these plans on television, the minister of transportation said that this would prevent the endless obstruction of progress by activists fighting for the protection of the environment. This was a political statement, in which "progress" was valued more highly than conservation.

The words of the minister are supported by a proposal by a Dutch employer's organisation to stop subsidising non-governmental organisations involved in the protection of the environment, because they misuse public money for actions and legal procedures against necessary construction projects. They also called for a boycott of the lottery that gives financial support to Greenpeace. Obviously, the employers do not understand that these actions and legal procedures are the only effective antidote to the propaganda of private and public enterprises carried out with money provided by customers and taxpayers.

Also in line with the minister's comments are the recent investigations by the popular press into the pasts of members of parliament and ministers. These politicians were active about thirty or forty years ago in action groups with social, pacifist, or "green" goals. As idealistic young people, they participated in massive demonstrations, radical actions, and other extra-parliamentary activities. This witch hunting can be compared with the McCarthy period in the USA, when everybody with leftist sympathies was boycotted. The words of the Dutch minister of transport and the proposal of the employers to terminate subsidies, on the other hand, are more like the reaction of a former English king, who did not like to hear unfavourable reports about the losses of his army on the battlefield. Messengers delivering such reports were put in jail.

It is unfortunate that massive demonstrations by idealistic young people, as we saw in the sixties, seventies, and eighties of the twentieth century, have become rare in the Netherlands. There is enough reason for young idealists to become active again, as the emerging scarcities and climatic changes pose grave threats to the future of the world. The only occasions where big crowds of young people gather seem to be sports events, pop concerts, and dance festivities. It reminds me of the dancing manias in the middle ages, which were a reaction to the uncontrollable threats of pests and diseases.

Real actions are left to professional organisations such as Greenpeace, Friends of the Earth, Amnesty International, and the World Wildlife Fund; but it seems that angry crowds can be raised only for materialistic goals such as wages, taxes, and prices.

17. The credit crisis *(13 October 2008)*

Although the present credit crisis was triggered by increases in prices of energy, raw materials, and food, its real cause is overconsumption and overproduction, particularly in the USA. People have been borrowing too much money, thinking that the economic growth would last forever. When they could not pay the interest for their mortgages and other debts anymore, the banks who lent them the money incurred financial problems too. Unfortunately, these banks, in turn, had been borrowing from other banks in other countries. That is how the crisis spread all over the world.

According to the capitalistic doctrine of the functioning of markets, the irresponsible individuals, firms, and banks went bankrupt. People with mortgages that were too high had to sell their houses, leaving them in debt. For young people with savings, instead of debts, this created opportunities to buy houses for much lower prices. Firms that had expanded too much were forced to reorganise or sell, which created opportunities for new, inventive enterprises.

So far, the capitalistic ideology worked out as it was meant to do. The levels of consumption and production decreased, and the cherished economic growth changed to economic decline. But now see what happens. The private banks, which borrowed too much from each other and lent too much to firms and individuals, get full protection from their governments.

Hundreds of billions of dollars, pounds, and euros are spent by governments to support these icons of capitalism. In some cases, this support consists of a downright nationalisation of banks, a move that had been considered a very detestable socialistic action in the prevailing neo-liberal ideology. The treasurers of the governments were in haste to declare that this nationalisation is only a temporary measure, as the shares would be sold again as soon as the situation in the financial world stabilises. In other cases, the governments guarantee the borrowing and lending between banks, in order to promote confidence and stimulate the flow of money. In both these cases, a great risk is taken at the expense of the common taxpayers, who may well have to pay the bill for overconsumption in rich countries.

If we consider the effects of these governmental actions, the only certainty is that a continuation of activities is guaranteed for the highly paid managers of banks, the irresponsible middlemen selling the loans, and the wizkids inventing the many complicated, tempting, and dangerous financial products. The victims of this "business as usual" will again be the honest and hardworking citizens, who will have to pay sooner or later for the interventions by their governments.

The least one can hope for is that all banks will henceforth be subjected to very stern regulations and supervision. Loans and mortgages should be given only under strict financial conditions and should be judged by their social and environmental effects. Complicated "financial products" should be forbidden. Bank managers should be held financially and ethically responsible for the actions of their organisations.

The extremely high salaries and bonuses for managers of banks and multinationals should be forbidden too. When the Dutch government nationalised the ABN AMRO Bank, it put a former attorney general in overall charge of the bank for a salary that was about ten times less than that of his predecessors. This illustrates that there is no relationship between the quality of managers and their salaries. Managers of banks, big companies, and semi-government organisations form an in-crowd, in which the maximization of personal incomes seems to be the main goal. It is not surprising that persons focussed so much on their own financial positions fail as managers. The big question is why institutional investors such as pension funds, which are holding most of the shares, are not stopping these excessive remunerations. Is it because they are represented in the meetings of shareholders by commissioners who will sooner or later be themselves eligible for a post as top manager? People who strive so much for excessive personal wealth are not fit for and worthy of being our leaders. By their bad example, they promote the present materialistic attitude of our world.

The Netherlands tried to stop the disproportional salaries and bonuses by voluntary agreements between big firms, but in practice, this was not followed by a change of attitude. Later on, there was talk of a so-called Balkenende-norm, named after the present Dutch prime minister. According to this norm, managers of semi-government organisations, such as the railways, postal services, universities, and hospitals, should not earn more than the prime minister earns. It is clear also that this norm can be imposed only by law, and that it should be applied also to private organisations. All these measures would suppress the extreme consumerism and "producerism," and might even help to stop the economic growth altogether. It might thus create the necessary conditions for a gradual stabilisation of the economy.

18. Economic stability *(21 October 2008)*

Economic growth is what economists and politicians have been advocating since long ago, in the communistic as well as in the capitalistic world. Economic decline, on the other hand, is feared like the plague in the middle ages. The official reason underlying this assumption is that economic growth will create the means to abolish poverty. Instead, economic growth has not prevented the rich from becoming richer and the poor from becoming poorer. The number of extremely poor people in the world is still rising steadily and has now reached almost one billion.

In between growth and decline lies stability. That is a situation in which the world is not aiming at an ever-increasing production and consumption, but at a gradual replacement of harmful production processes and products by beneficial ones. The expected recession following the present credit crisis may well create the circumstances to do just that. Fossil energy sources can be replaced by renewable and clean energy. Plastics and other materials derived from oil should disappear in favour of materials that can be decomposed or recycled. Manures should be used instead of artificial fertilizers. Cars and machines powered by oil must give way to electric cars and machinery. The highly polluting traffic by aeroplanes and boats must be replaced as much as possible with transport by electric trains, busses, and cars.

All these changes will cause a loss of jobs in the conventional production sectors, but at the same time, will create challenging new job opportunities in the innovative sphere. In Germany, where 15 percent of the energy requirements is already provided by renewable resources, the production and installation of solar panels is already providing more jobs than the production of conventional cars. Also in Denmark, renewable resources provide already about 20 percent of the total energy consumption. There, the manufacturing of windmills is a booming business.

If by chance these changes resulted in a net loss of jobs, work and incomes should be distributed more evenly among workers. This can be done by shortening the workday or workweek. In that way, the loss of income is compensated by a gain of leisure time. After all, people do not live for work but work to live.

To bring about these changes, governments can use tools such as laws, regulations, subsidies, and taxes. This must be coordinated on a worldwide scale, to prevent false competition between countries and firms. It is clear that this cannot be done under the prevailing neo-liberal ideology. Governments and international organisations such as the United Nations and the World Bank must take their responsibilities to lead the world to a better future.

19. Consumerism and producerism *(29 October 2008)*

There once was a time in the Netherlands when the birth of a baby was celebrated with sugared caraway seeds on rusks. Visitors would bring some clothes for the baby, often personally knitted. Birthdays were commemorated with congratulations, coffee with cake, and small gifts. During the Dutch Sinterklaas Festival, some sweets and presents were given to the children and funny poems were written for the grownups. Christmas was mainly a family reunion with a tasty meal cooked by the hostess, and New Years Eve was celebrated with some fireworks and homemade doughnuts.

Nowadays, all these occasions are embraced as an opportunity for endless shopping. Particularly during the Sinterklaas Festival in December, huge amounts of mainly plastic toys for the children and often useless presents for the grownups are sold in the shops. To create even more buying and selling opportunities, the Dutch have also adopted the Anglo-Saxon habit of offering presents under the Christmas tree. The third opportunity for big spending in December is New Year's Eve, with its ever-increasing display of fireworks.

This report was not written to suggest that the Dutch are the biggest spenders in the world, as we know that this type of overconsumption is taking place in all rich countries. The only hope for an end to this "consumerism" is based on the expectation that an economic crisis will lead to a regression or even a lasting depression.

Unfortunately, governments are still trying frantically to prevent this economic decline, as is shown by their financial support of banks and industry. They are strongly supported in this effort by all sorts of public organisations. A Dutch labour union has even suggested that compulsory savings by employees, set aside for pensions or other important future goals, should be freed for shopping in December 2008. This is just one example to illustrate how the world is still aiming at a continuation of consumerism and producerism.

One explanation for the current materialistic attitude in the rich countries may be that only the oldest generations have experienced wars and other periods of scarcity, but other generations have not. The younger people think that wealth is a normal condition, without realising that it is based on the destruction of their environment and on the natural and human resources of countries in other parts of the world. This can also explain why people in the rich countries waste and throw away such huge amounts of food.

The World Wildlife Fund published a report today, dealing with the natural resources available to humanity. It is expected that by 2035, the human species will need twice the surface area of our planet to support itself on a sustainable basis. As we have only one planet earth, this will, without doubt, lead to a continuing destruction of the environment and depletion of resources.

20. Minerals *(13 November 2008)*

It is clear that renewable energy, in the long run, can cover almost all the energy requirements of the world. The same is true for bulk materials used for building and construction, such as cement, iron, and aluminum. Huge amounts of these are available in nature in the form of limestone, laterite, and bauxite. The recent price increases of iron and aluminum were not due to a lack of availability of these raw materials, but to the fact that the processing of them could not keep pace with the demand in rapidly developing countries. Still, it is important to keep in mind that the mining, processing, and transport of these materials require much energy, destroy landscapes, and pollute rivers and the atmosphere. Therefore, iron and aluminum must be recycled as much as possible.

Other useful minerals are not available in such abundance, and structural shortages of them will occur in an ever-growing world economy. Examples of such minerals are copper, zinc, manganese, phosphate, silver, gold, and diamond. Nations are already manoeuvring to safeguard their future requirements of these materials, which leads to serious international conflicts. An example of such a conflict is the recent invasion of the Tutsi militia under General Laurent Nkunda into Kivu in Eastern Congo. Officially, the objective of this invasion was to protect the interests of the Tutsi minority in Eastern Congo against the Hutu majority. The real purpose was obviously to get hold of the rich geological reserves of copper, cobalt, and coltan. The latter is used for the production of mobile telephones and computers. Already, these minerals, which were exploited earlier by Hutu militia, are exported via Rwanda under the protection of General Nkunda. In return, the warlord receives modern weapons and ample money to pay his soldiers.

It is not difficult to guess from where the money and the weapons are obtained, as modern weapons, computers, and mobile telephones are manufactured only in rich countries. In fact, the Chinese government has signed an agreement with the Congolese government for additional resources. Under this agreement, the Chinese will invest eight billion US dollars in the improvement of the Congolese economy. The exploitation of mines in different parts of Congo is a part of this agreement. Possibly the improvement of the morale and outfitting of the weak Congolese army in Kivu is a less publicised ingredient of the deal. Also, Russia has recently delivered combat helicopters and aeroplanes in Kinshasa, in exchange for mining concessions in the province Katanga.

To prevent further tensions, scarce minerals must be recycled on a large scale. This would help to keep a stable world economy running without international conflicts. Further reduction of the economic growth is an essential ingredient for this.

Water is an essential commodity for many human activities. In nature, it is recycled on a large scale by evaporation and precipitation. Humans can use it for their own purposes by intercepting and conserving it during this cycle. Unfortunately, the water is not always available in all places in the desired quantities. Therefore, human activities must be adapted to shortages or excesses of water.

In deserts and semi-deserts, precipitation is insufficient to support agricultural crops. Rainfall is irregular in time and space. Only nomadic animal husbandry is possible. In the past, nomads have established very intricate grazing routes, to make full use of the production capacities of different locations. In that way, they were adapted to the climatic conditions in which they lived. Later on, modern techniques made it possible to dig deep wells for the exploitation of groundwater. The result of this has been a deterioration of the nomadic grazing systems and a concentration of cattle around these water sources, causing a destruction of the natural vegetation around the wells. In many places, even this situation is not durable, due to the depletion of fossil groundwater resources.

Near the deserts and semi-deserts, one finds semi-arid and sub-humid regions, where it is possible to grow only one crop per year. The farmers in these areas were always dependent on the irregular occurrence of rain showers. When the rainfall failed, people would starve and die, which kept the population density in equilibrium with the low production capacity of the area. More recently, food aid from elsewhere during meagre years has caused an ever-growing overpopulation in these regions.

One way to grow more crops in areas with insufficient rainfall is to import water. This can be done by tapping it from rivers carrying water from elsewhere. In recent times, dams have been built in all dry parts of the world, to direct river water to irrigation canals, and from there to agricultural fields. In that way, agricultural production of a country or region can be increased considerably. Due to this, there is already a strong competition for water, particularly where the source rivers cross international boundaries. Within India, many disputes have taken place between different states around the same river, but these could be settled peacefully with the help of the federal government.

In many situations, also irrigation is not a durable system. If the amount of irrigation water directed to an area is too high in comparison to the natural drainage capacity of that area, water logging and salinity will occur in the long run. This has caused ancient civilisations in Mesopotamia, the Indus plains, and other riverine areas to disappear, and is now affecting many newly irrigated areas in India, Pakistan, and other countries.

If the rivers feeding the irrigation systems carry much silt, artificial lakes at the dam sites will silt up and lose their potential to store water for dry periods. This happens particularly where natural vegetation is destroyed in overpopulated parts of the catchment areas of these rivers. In Northern India and Southeast Asia, but also in other regions where rivers are fed mainly by the annual melting of glaciers, another danger is emerging. If these glaciers melt due to global warming, a constant and dependable supply of clear water to the rivers will disappear. Instead, there will be widespread flooding in the lowlands during rainy seasons and shortages of water during dry periods.

Large-scale irrigation projects have several ecological disadvantages too. The water reservoirs upstream of the dam sites often submerge hundreds of kilometres of beautiful and ecologically valuable riverine areas, harbouring prosperous villages of farmers and fishermen, monuments, and even towns. This is the reason why the local inhabitants of the valley of the once mighty Armada river in India protested vehemently against the construction of a dam for irrigation purposes. In China, the same destruction of landscapes and habitations is happening due to the construction of a "mega-dam" in the Yangtse Kiang River. Downstream of the dam sites, ecological disasters are occurring as well. Near Madurai in South India, houses have been built in the completely dry bed of the once beautiful Cauvery River, because a dam upstream is diverting all the water to irrigation areas.

The introduction of modern irrigation is also not favourable for the local landscape and its inhabitants. The original vegetation and cultivation patterns, following natural variations of the landscapes, are replaced by linear irrigation canals and channels with high bunds around rectangular irrigated fields. The natural vegetation is often destroyed completely to give way for agriculture, and the original local farmers generally lose their land and become agricultural labourers for the new landowners.

The increase of agricultural production by destructive methods should be stopped. Instead of constantly trying to increase the amount of food on our global table, it is time to stop increasing the number of guests at the table.

22. Fish *(30 November 2008)*

Fish has always been a part of the human diet. Even in prehistoric times, fishermen tried to catch fish in rivers, lakes, and seas. Fishery could be carried out on a sustained basis as long as the world population consisted of small communities, but with the recent strong growth of the population, the fishing industry has been catching more fish than the seas can supply on a sustainable level. Moreover, industrialised fishing methods have become so aggressive that they destroy the habitats of marine live and catch inedible species such as ray, shark, dolphin, seal, and porpoise as discards.

As a result of these developments, some species of fish have become almost extinct. The populations of tuna and swordfish have decreased worldwide by as much as 90 percent and the availability of cod in the North Sea has been reduced in fifty years to far below its sustainable level, from three hundred thousand to fifty thousand tons. The amount of salmon in seas and rivers has been reduced to one quarter of its original magnitude. Approximately one third of the edible fish species in the North Sea have been overfished. Amongst these species are, apart from those mentioned earlier, cod, eel, plaice, turbot, sole, whiting, seawolf, and seadevil.

Fifteen years ago, the European community agreed that one quarter of the North Sea would be closed for fishing. Most governments have not yet earmarked the locations of these sea reserves. The Dutch minister of fisheries declared recently that further studies are required for the selection of those areas. In the meantime, European fishermen have been shifting their activities to the coasts of Africa, where they are destroying the natural fishing grounds of their African colleagues with their rough dragnets.

It is clear that the European level of fish consumption cannot be introduced in the whole world if the present high growth of the world population is continued.

At present, almost half of the world population is living in towns. Particularly in the developing countries, the big towns are growing fast. The new town dwellers are people from the countryside, in search of a better standard of living. This massive migration to the towns is caused mainly by the fact that the villages and farms offer insufficient job opportunities for the growing population. Due to the high population density, farms and firms are split up until they become too small for the maintenance of a family. Hence, some of the sons and daughters of the farmers and villagers have to try their luck elsewhere. In the towns, the new migrants settle into the improvised shelters of sprawling shantytowns. Most of them manage to earn a meagre income as daily labourers, and a lucky few find a permanent job. The shantytowns are still growing fast and this development will go on as long as the population growth continues.

Many people from the rural areas and shantytowns in the poor countries try to migrate also to the rich nations. The wealthy countries try to stop them by all means, but some manage to get through the political and physical barriers. In some countries, the people in overpopulated rural regions migrate not only to the towns, but also to other rural areas. An example of this is the so-called transmigration of Javanese people to other Indonesian islands, notably Sumatra,

Kalimantan, and West Irian. This movement was initiated by the Indonesian government shortly after independence. In their new environment, the migrants get the right to reclaim land for agricultural activities. Many square kilometres of forest were lost in this way, and the transmigrations are still continuing.

In Brazil, a similar movement is happening. People from the poor, overpopulated areas of that country are migrating to the Amazon region. Along newly built roads, they reclaim small pieces of land for their own use. It is one of the developments leading to a gradual disappearance of the tropical forest in Brazil.

Overpopulation occurs in the economically developed world too, but there the cause is overproduction and overconsumption in combination with a high population density. Due to these conditions, much land is used for industrial areas, roads, housing, recreation facilities, and some nature reserves. The pressure on the land would be released there if a more moderate lifestyle would be adopted. After all, unhappiness is more often caused by poverty or extreme wealth than by moderation. It is, therefore, reasonable to expect that rich countries stop the growth of their economies.

24. Nature *(30 November 2008)*

The growth of the world economy and the world population leads to a strong pressure on nature. Legal and illegal cutting of trees occurs on a large scale, particularly in Brazil, Indonesia, Kampuchea, Cameroun, and Russia. In Argentina, the cultivation of gentech soja for cattle fodder covered an area of six million hectares in 1995 and 14.2 million hectares in 2003. This means that an area the size of The Netherlands was deforested in a period of eight years. In Brazil, even more tropical forests are disappearing, due to the cultivation of soja, the development of grazing land, the growth of sugarcane for the production of bio-fuel, and the illegal settlement of migrants from the northern part of the country. In Indonesia, transmigration, the illegal cutting of trees, and the production of palm oil are the main causes of the disappearance of forests. Approximately 50 percent of the tropical wood imported in the Netherlands is cut illegally.

The loss of natural landscapes, both on land and at sea, leads to a disappearance of plant and animal species too. According to the "Living Planet Index", published by the World Wildlife Fund, there was an average reduction in size by 27 percent for four thousand populations of a few hundred land species of animals between 1970 and 2005. At sea, there was a 28 percent decrease on average of the size of the populations of large fishes between 1995 and 2000.

The International Union for the Conservation of Nature (IUCN) presented, during its latest annual congress at Barcelona, a report on the condition of 5,487 species of mammals. Half of the species are decreasing in the number of individuals and one quarter of the species is endangered. The Orangutan belongs to the endangered mammals. It would be a disgrace for humanity if it were responsible for the extinction of one of its closest relatives.

25. Basic instincts *(16 December 2008)*

The success of reproduction of an animal or plant species has always been determined by its ability to adapt to different circumstances. The ancestors of the human species could survive changes of the climate and other natural conditions when they learned how to use their hands for purposes other than climbing trees and picking fruit. Gradual changes in the brain made it possible to apply the use of hands for many different activities, such as the manufacturing and use of tools. Thanks to these abilities, the human species became very successful in its reproduction. It could adapt to the most extreme climatic conditions and, thus, could spread to all corners of the world. Mankind could even create its own microclimate, as it was able to make clothes, build shelters, and control fire. With their manipulative hands and manipulative brains, humans could not only adapt to many different environments, but also change the environments to their own requirements. No wonder that the urge to manipulate became one of the main basic instincts of the human species. It is one of the main causes of producerism.

The urge to consume is a characteristic of all animal species. Food is always consumed in big quantities when available. Apart from physical effects on the body, a safe supply of food had important psychological and social consequences for individuals of animals living in groups. Well-fed individuals were strong and could attain power and self-confidence. With their strong bodies, they could impress and control the other animals in the community and, thus, could become leaders. Humans attain self-esteem and power also by the possession of non-edible commodities such as clothes, big cars and houses, ornaments, and money. This is the cause of consumerism. In fact, money has become the most important commodity in this respect, because with money, one can buy not only all the other material things, but also such immaterial matters as security, comfort, respect, power, and sex.

In the beginning, all these activities could be carried out on a sustainable basis. Even agriculture and animal husbandry were not really harmful for the environment, as there was enough space left for the conservation of natural resources and the survival of animal and plant species. Problems arose when the human population started growing explosively and the production of consumable goods was taken up on an industrial basis. In the preceding parts of this blog, these problems were described in some details. Energy, water, minerals, nature, and space are becoming scarce commodities and the conditions in the atmosphere, on the land, and in the seas are changing at a harmful and frightening rate.

The present worldwide credit crisis may force the human population to decrease production and consumption temporarily, but it is clear that great efforts are being made all over the world to resume economic growth as soon as possible. It seems psychologically impossible for humans to reduce their production and consumption activities to an acceptable level. Therefore, the most important contribution to a sound future of mankind may be to stop the growth of the world population. This will be difficult to achieve too, as the reproduction instinct is enshrined in the human mind as a natural right.

People in power have always used this human urge for their own purposes. Worldly rulers needed soldiers for their armies, factory owners required workers for their factories and spiritual leaders wanted large crowds of followers. Even now, the Roman Catholic Church and other religious institutions are opposing the use of family planning. The only spiritual leader advocating birth control nowadays is the dalai lama of Tibet.

Humans have always boasted that they are the only rational and ethically motivated creatures on this earth. These faculties should be used now to control their own basic instincts for production, consumption, and reproduction. Let us stop growth now, in the interest of our descendants and to save the beauty of our world.

PART II
THE POSTSCRIPTS

1. Recent developments *(9 July 2009)*

Half a year after the preceding blog was completed, there are clear signs that the awareness of certain problems is growing. In particular, the threat of climatic changes due to the use of fossil energy is more widely recognised. The election of a new president in the USA has been an important factor in this change of attitude. For the first time, the USA is siding with Europe in an effort to cut down the use of fossil energy and replace it as much as possible with sustainable energy sources.

The present financial and economic crisis has been helpful in the reduction of the use of fossil energy. Big cars are being replaced by small cars, air traffic is decreasing, and many favourable technical innovations are taking place. The government of Great Britain has recently published a plan in which technical measures will be used to reduce CO_2 emission by 34 percent of its present magnitude by the year 2020. For Europe, there are even proposals to create a continental electrical network for the distribution of solar energy from the Sahara and Arabic countries, hydro-electric power produced in mountainous European regions, and aero-electric energy from windmills at sea.

Unfortunately, all these developments will be insufficient to meet the energy requirements of the developed countries at the present level of consumption. The European plan will take hundreds of billions of euros and several decades of time to be realised, and will then supply only 15 percent of the expected European energy needs. Under these conditions, the call for the use of dangerous nuclear energy is rising. In the Netherlands, a request has already been filed for the construction of a second nuclear plant, to become operational in 2018, which will supply only 2 percent of the country's electricity requirement. In Germany, where the majority of the population has been opposing the construction of new nuclear plants for a long time, and where already 15 percent of the energy used is supplied by solar panels and windmills, the production of nuclear energy is again on the agenda.

Enough has been written in the preceding blogs about the dangers of the production of nuclear energy. A clear illustration of these dangers is a recent occurrence at Asse in Germany, where radioactive waste from different European countries is stored in former salt mines. The barrels containing nuclear waste were being corroded by salt water leaking into the mines, which caused the contamination of groundwater with radioactivity. This is a good example of the dangers of unexpected and unpredictable whims of nature.

In the meantime, the shortages of basic materials such as food are still increasing. The number of people suffering from lack of food has surpassed one billion in 2009. It is expected by the United Nations that the demand for food will have doubled by the middle of this century.

Despite the temporary economic decline in the developed countries, the world economy as a whole is still growing. This is due to the continuing expansion of the new emerging economies. The economy of China has shown an annual growth of 7.9 percent in the second quarter of 2009. All countries are still trying to end the present economic recession and to resume economic growth as soon as possible. This will certainly lead us again to worldwide shortages of oil and other basic commodities.

2. The G8 again *(12 July 2009)*

Again, the eight countries with the largest economies of the world held their annual meeting. This time they did not ask the oil-producing countries to increase their production. Thanks to the recession, oil prices have come down drastically since July 2008. There is no danger anymore that the worldwide oil reserves will shrink to an unacceptably low level. Instead, fully loaded tankers are parked outside harbours, waiting for an increase in the oil price. As economists are expecting that the recession will continue for several years, the tankers may have to wait there for a long time.

Again, the G8 promised to donate and invest billions of dollars to fight poverty and hunger in the poor countries. It is doubtful whether this aid will come forward, as also in previous years, the help given has remained far below the promises made. In 1970, the rich countries agreed to donate 0.7 percent of their national incomes to development aid, but up to last year, they have reserved only 0.3 percent of this income for this purpose. The United Nations Organisation and the development organisation Oxfam have urged the G8 to raise their contributions, but due to the recession, few countries are inclined to increase foreign aid. The UN and Oxfam have also recommended replacing direct food aid with agricultural development projects. However, this proposal has little chance to be implemented, as food aid is used by the rich countries to support their own agricultural production.

Instead of investing money in the development of the local agriculture in poor countries, nations with rapidly expanding economies such as China, India, Qatar, and Saudi Arabia are now buying and leasing large areas of agricultural land in Africa. Due to the emerging global food shortages, these developing countries will need the land in the near future to produce enough food for their own populations. The G8 have not called for a termination of this new trend. They only recommended that good principles be established for these practices.

The most important item on the agenda of the G8 was the climatic changes. All countries showed awareness of the magnitude of this problem, which resulted in far reaching intentions. They agreed that the average global temperature should not rise more than two degrees above the temperature at the beginning of the industrial revolution, as a further increase would have very serious implications for the world. This agreement implies that the worldwide emission of CO_2 must be reduced by 80 percent by 2050 and that it must reach its maximum in 2015. Mr. Ban Kimoon, the secretary general of the United Nations, commented that even this reduction is not sufficient to stay within a 2 percent rise in temperature, and the International Energy Agency stated that the expenses for a complete change to renewable energy sources by 2050 would be around four hundred billion US dollars per year.

The Western countries have proposed that the expanding economies of China, India, Brazil, South Africa, and Mexico will reduce their CO_2 emission 50 percent by 2050. These so-called G5 countries are not keen to follow this recommendation. In their view, the rich countries should take concrete measures first. This is a reasonable demand, as long as the recommendations of the G8 have not been followed by clear actions in the rich countries.

3. Fishy business *(20 July 2009)*

A kind of tuna fish, called blue-fin tuna in the Netherlands, is one of the largest fishes found along the coasts of Western Europe and in the Mediterranean Sea. There, it is caught by fishermen from Spain, France, Italy, and other European countries. It is used on a large scale for the production of sushi in Japan and in the Western world. Environmental organisations have warned for years that this tuna species is on the brink of extinction. Finally, Monaco, France, the United Kingdom, and the Netherlands have decided to stop the fishing, trade, and consumption of this fish.

A consequence of this decision is that sushi producers will start using, and fishermen all over the world will start catching, other species of tuna. Thus, the pressure on the populations of blue-fin tuna will be diverted to other tuna species. This diversion of pressure is also taking place with other species of fish. In the Netherlands, a so-called "fish guide" has been published, in which the species of fish used for consumption are classified according to the seriousness of the danger of extinction. In as far as this fish guide is used by the consumers, this danger will be shifted from one species to another. Only a reduction of total fish consumption, which can be brought about by a reduction of the world population, can bring an end to this alarming situation.

Also related to the problem of overfishing the seas are the recent reports of a strongly increased activity of pirates in the Gulf of Aden. Most of these pirates are former fishermen, who have lost their fishing opportunities due to the recent activity of large European fishing boats in their traditional fishing grounds.

An even bigger threat to the supply of healthy fish in the long run is the rapidly increasing pollution of the seas. Particularly, the accumulation of very-slowly-decomposing plastics is alarming. In the Pacific Ocean, an area has been discovered where these synthetic materials are concentrated by circulating currents. This area, called the Great Garbage Patch, covers an area twice the size of the American state Texas. The depth of the "plastic soup" is estimated to be many metres. According to the United Nations, who held a "World Oceans Day" recently, similar accumulations exist in five other locations. The North Sea alone receives twenty thousand metric tons of garbage annually. Most of the plastics decompose very slowly, but they gradually fall apart in microscopic small particles, the so-called micro-plastics. These micro-plastics are consumed even by plankton, which is the base of the maritime food chain. Hence, the micro-plastics also enter the human body via fish consumption.

4. Food supply *(22 July 2009)*

The United Nations has reported that approximately half of the world population is now living in towns. In poor countries, these towns are still growing fast. Some of them have already reached tens of millions of inhabitants. All these town dwellers are deprived of space, natural views, daily quietness, nightly darkness, and nature. That is why they try to escape, at least once every year, to crowded holiday resorts by means of crowded roads, aeroplanes, or trains.

Even worse is that townspeople do not produce their own food. They depend completely on the food production in rural areas for their food supply, which causes massive transport over large distances. Moreover, they want to buy this food for the best possible quality and the lowest possible price. Supermarkets and smaller shops are constantly competing with each other to satisfy these wishes of their customers. As a result of this, the producers move to the most profitable production methods. This involves the introduction of ever-larger production units and of production practices that ignore the interests of the environment, nature, and production animals.

In the Netherlands, and most other countries, mega-farms exist already for the fattening up of pigs, chickens, and other animals. To keep the production costs low, chickens are kept by the tens of thousands and sometimes more than a hundred thousand in one huge production unit. To further decrease production costs, the number of animals per square metre is maximized; for the production of eggs, chickens are kept in small cages with up to fifteen individuals per square metre.

In the Netherlands, there has been, for many years, a hallmark for food that is produced under strict ecological conditions, which includes the way in which farm animals are kept and treated. Naturally, food with this so-called EKO-mark is considerably more expensive than food produced in "regular" farms. Due to this, these products have conquered a share of the market of not more than 5 percent. It is obvious that people in general are not prepared to pay for the protection of the environment and for the living conditions of animals, even in a rich country such as the Netherlands. Therefore, governments should either subsidise the products of ecological farming or impose an eco-tax on the products of "regular" farming. So far, the Dutch government is not prepared to do this. It has taken the stand that ecological products must conquer the market by their own strength, just like other products. This is a typical neo-liberal attitude, which ignores moral, ethical, and aesthetical considerations.

Not only in animal husbandry, but also in horticulture, there is a movement towards development of mega-farms at the expense of the environment. A good example is present around Lake Navasha, which once was a beautiful natural lake in Kenya. There, a massive development has taken place towards the production of roses. This industry could be established there due to the availability of masses of poor people, who can be employed for low wages and under bad working conditions. The lake level is receding due to the use of water in the greenhouses and the water in the lake is polluted by the continuous release of pesticides.

These developments towards ever larger production units with destructive side effects could be retarded if the movement of people to the cities, and hence the growth of human populations, would be stopped.

5. Wishful thinking still *(25 July 2009)*

Still many people think that the climatic change is not to be taken seriously. They are of the opinion that this is a hype triggered by Al Gore. Mistakes made by Al Gore in his famous film *An Inconvenient Truth* offer them an opportunity to reject the whole idea of climatic changes. They forget that Al Gore's documentary is just a journalistic product. The real message comes from the International Panel on Climatic Change (IPCC), a body of hundreds of climatologists functioning under the auspices of the United Nations.

Some authors base their criticism on the data published by the IPCC itself, but they forget to update their information with the latest findings of the panel. One of them is Bjorn Lomborg, teacher at the Consensis Centre at Copenhagen, who still thinks that the sea level will rise by not more than thirty centimetres during this century. Although even this rise is disquieting enough for inhabitants of low-lying deltas and islands, it is clear that the climatic change and the rise of the sea level are taking place at a much faster rate than the IPCC estimated some years ago. In the meantime, the IPCC has been refining its models and has admitted that the changes are going faster than they had originally forecasted.

Bjorn Lomborg has drawn the conclusion from his observations that we should not talk too much about the emerging climatic changes and rise of the sea level, because it is discouraging for our children and grandchildren. They might loose confidence in the future and have bad dreams about catastrophes. If we were to follow Bjorn Lomborg's advice, the catastrophes might come without us even mentioning them. Would that be in the interest of our descendants?

Real scientific criticisms on the forecasts by the IPCC and its members are found in the book *The Deniers*, edited by Lawrence Solomon, in which scientists of different disciplines have written chapters. About half of the contributions in the book deal with details and do not really undermine the validity of the main conclusions of the IPCC. Other contributors claim that the models of the IPCC are still inadequate to produce detailed forecasts of global warming and climatic changes, a point that the IPCC itself recognises by refining its models all the time.

6. Prospects for the future *(29 July 2009)*

I t was remarked in the first postscript that the rising awareness of the problems ahead of us has resulted already in some concrete actions. These actions are mainly in the technical sphere. Examples are the development of small cars with low petrol consumption and cars powered by hydrogen or electricity. Better isolation of buildings and the introduction of solar panels and windmills for the production of electricity are other examples. Many economists and politicians think that a recession is the best condition in which to start such developments. Old production lines, rendered out of date by recent developments, will go bankrupt, and governments must try to ease the resulting unemployment by stimulating new activities. This creates opportunities for the development of new products and production processes, directed towards the solutions of expected global problems.

It is clear that the expected problems cannot be solved by technical innovations alone. The main causes of the climatic changes and the shortages of basic commodities are in fact overconsumption and overproduction. Our globe just cannot continue to fulfil all our increasing material wishes on a sustainable basis and, in the long run, all technical "solutions" will be of no avail. What is needed is a complete change of the political, economical, and social character of our human civilisation. The present financial and economical crisis offers a unique opportunity to bring about this change. A result of the crisis will be that citizens will face a reduction of their personal incomes. Thanks to this, people will get used to lower consumption levels. At the same time, production will go down due to the bankruptcy of factories and the laying off of workers. If this reduction of consumption and production could be maintained beyond the end of the economic crisis, a new human civilisation can take shape. This development will not be a new socialistic revolution or a restoration of neo-liberalism, but rather what I would call a movement towards a moral society.

First of all, it is important that the burden of the new situation not be loaded on the shoulders of unemployed people. In principle, there is a solution for the expected unemployment of around 10 percent. If the working hours of all employees would be reduced by 10 percent, unemployment would not exist. The result of this measure would be a reduction of the production, personal incomes, and consumption by 10 percent. If this system would be made permanent in all countries, the global economy would decline by 10 percent on a sustainable basis.

Another possibility to reduce production and consumption is the gradual replacement of energy-intensive production methods by labour-intensive methods. This mechanism would automatically become active when the present economic crisis ends, thanks to renewed increases of the scarcity and price of oil. Governments can support this mechanism by increasing the taxation of energy and reducing the taxes on labour. The result will be a reduction of labour productivity, production, personal incomes, and consumption.

There should be strongly progressive taxes on the incomes of individuals and the profits of companies. This would keep the essential system of competition and the functioning of markets intact and at the same time provide governments with enough money for essential communal provisions, such as free basic health care, free education, and financial support for the weakest members of the society. Teachers and preachers must play a role in this development by advocating that the highest goal in life is not the individual accumulation of wealth and maximization of incomes, but rather the development of talents, the attainment of knowledge, and respect for fellow citizens, nature, culture, and the environment.

There should be a strong supervision on the functioning of private banks and companies, to stimulate competition and to prevent the concentration of power and wealth in a few private hands. So far, Western governments have been able to influence the management of private enterprises only after giving financial support. This influence must be extended to all private firms, to make sure that they function for the benefit of the whole society and not only in the financial interest of their investors, employees, and managers.

I recognise that the proposals in this postscript are very basic. To implement them, many problems must be solved by economists, politicians, and diplomats. Particularly in democratic nations, much work has to be done to reach agreements between the governments, the employers, and the labour unions, and to develop suitable legislation. In the global sphere, there will be the obstacle of international competition, which will have to be regulated in worldwide conventions. This can only be realised as soon as the world population is sufficiently aware of the enormous dangers of overpopulation, overconsumption, and overproduction. Let us hope that this awareness will not come too late.